I0493970

Introduction

When people think of gardening, they usually picture it as an outdoor activity reserved for those with yards or access to an outside area. Gardening is an activity that is good for both you and your environment; it is relaxing and leaves you with a sense of achievement. With people increasingly living in apartments and in urban areas, they often assume that a garden is not even an option for them. The aforementioned assumption is incorrect because as this book will show you, an indoor garden is possible and can be a wonderful addition to your home and life.

Now that you are aware of the fact that you can have an indoor garden it is time to figure out where and how you can create the perfect garden. Whether you have a small space or a larger area this book can be your guide to success in the area of gardening. Many people make the mistake of starting an indoor garden without knowing all the ins and outs. Do not make the same mistake! "Indoor Gardening Made Easy" will assist you in all your indoor gardening needs.

One of the most important parts that will determine the success or failure of your indoor garden is the location, this book will give you the necessary information so that you can pick the perfect spot for your garden to thrive. Another area this book will cover is learning how to grow vegetables indoors, which will allow you to make meals with your home grown veggies! This eBook really goes in depth and gives you all the details you will need to make your indoor garden a grand success; from choosing the correct lighting to learning how to water your garden and much more!

There's no need to miss out on the joy and health benefits gardening brings just because you may not have an outside space. Do not hesitate, whether you live in an urban loft, a studio apartment, or a traditional house, "Indoor Gardening Made Easy" will be your guide to bringing a little bit of outside, inside.

Contents

Chapter 1

The Benefits of Indoor Gardening.

Chapter 2

Choosing the Ideal Location for an Indoor Garden

Chapter 3

How to Select the Ideal Lights for Your Indoor Garden

Chapter 4

Selecting the Right Containers for Your Crops

Chapter 5

The Right Gardening Tools

Chapter 6

The Right Fertilizers

Chapter 7

The Growing Medium

Chapter 8

How to Water Your Indoor Garden

Chapter 9

Hydroponic Basics

Chapter 10

How to Grow Vegetables Indoors

Chapter 11

How to Care For Herbs Grown Indoors

Chapter 1

The Benefits of Indoor Gardening

Green urban living is a lifestyle in which people living in an urban neighbourhood or any other type of built up area can be as earth friendly and self-sufficient as possible even in their small space. By incorporating an indoor garden, you can live in a greener and more natural life in the comfort of your own home or apartment by using indoor plants and indoor gardening. Apart for the benefits of a living a greener lifestyle, indoor gardening has a few additional benefits as opposed to outdoor gardening including:

Greater control over the temperature and environment

When you grow vegetables or any other indoor plant, you have full control of the temperature of the garden. Some herbs or vegetables won't do well in a cold or windy environment, or an environment which lacks adequate sunlight. By growing an indoor garden you can control all these factors, which mean you can keep your plants safe from excessive wind or rain and allow for adequate sunlight.

You can protect your crops from pests

It's no secret that humans aren't the only ones who like to eat vegetables and herbs; a wide variety of pests including many small insects like to eat those same things. As well as insects, many other pests such as moles like to burrow, literally ruining your crops from the ground up. When you garden indoors this problem is eliminated.

You can protect your crops from parasites

Aside from pests, when you plant your garden outdoors you also have to contend with various parasites which can attach themselves to your crops. By gardening indoors, many types of outdoor parasite problems can be eliminated. However, you will still need to be vigilant as a minority of parasites can find their way indoors.

You can enjoy the benefits of a greener life style which include:

Better indoor air quality

House plants can improve your indoor air quality be absorbing hydrocarbons from furniture and detergents, chemicals from building materials and filter allergens from the air.

Within twenty-four hours they can remove up to eighty-five percent of toxins from the air and in return precious life giving oxygen. The toxins they can absorb include formaldehyde, acetone, benzene, carbon monoxide and trichloroethane which can be emitted by adhesives, paint, varnishes and caulking compounds.

Therapeutic Benefits

•Recent scientific studies have proven that gardening can lower blood pressure, reduce depression, and promote healing.

•Indoor gardens can increase the humidity levels for a cleaner more oxygen-rich environment.

•An indoor garden can add beauty to your home.

Bonus benefits include:

•**A Convenient source of fresh food;** what could be more convenient or fresher then walking over to your basil plant and picking a few leaves for your favourite pasta dish?

•**A great educational source for children;** what better way to teach children a valuable life skill than to have them grow their own food?

•**Personal satisfaction;** Growing your own food can be a very rewarding and satisfying experience.

Chapter 2

Choosing the Ideal Location for an Indoor Garden

Real estate agents will tell you that the three most important factors when determining the value of your home is location, location, location. The same statement is true when planning your indoor garden, whether you intend on starting a small-scale garden or a commercial enterprise. The right spot where to locate your indoor garden is an important consideration for a few reasons, both from a practical and an aesthetic view point. First, you are going to spend a lot of time working on this garden, nurturing your crop until it is ready to harvest, so you want it to be positioned in such a way that it is easy to get to and manoeuvre.

Second, you will be inverting some money in this set-up; therefore you want your investment to bear fruit. It will take on average three to four months before your crops will be ready to harvest. So you first question should be: Can I can I still use this position in an freely and in an uninterrupted way until then? If you cannot, you should try to find somewhere else to put your garden. Once you crops start growing you may want to start a second crop, to maintain a vegetative growth phase until your first crop is ready to harvest. By doing this you will be able to harvest more frequently; say every two months instead of every four months, for example. You may eventually decide to have one garden for plants large enough to begin the bloom cycle when your first crop has reached maturity. This would mean having an separate area for both the grow and bloom plants, though the vegetative growing area is usually smaller in size. Where you position your indoor garden will also affect it's over all efficiency, including how much electricity it will consume.

If your garden is to be maintained in close proximity to living areas or neighbouring properties, the lights and sound of your growing position could also pose a problem. Basically, you're looking for a place which will provide you with the most convenient and trouble free growing experience, and several other factors need to be considered including:

Noise

Almost all indoor gardens have some level of noise. If you have neighbours living below or above you, this could be a problem.

You have to consider how much noise your equipment will make which will mainly depend on the type of growing equipment you install. If you decide to use high intensity discharge or HID lights, these will require a lot of cooling, which could mean having lots of air conditioners and fans or sometimes both. If you decide to use Fluorescent or LED lights, these are not as hot, so cooling requirements and therefore noise can be reduced.

In extreme circumstances you may consider building a stealth garden where almost no noise is created, this however, could compromise the overall performance of your growing equipment.

You may also consider acoustic tiling, which will leave a space between the sound and adjoining walls, so the sound is dampened and not easily transferred to other areas. As an alternative, you could use gym mats or convoluted foam to dampen the sound, just be sure to cover the floors with durable material which extends a few inches above the baseboard. This will prevent water damage to your floors and water leaking into your neighbour's property below.

You could possibly construct a room which could house a commercial scale Controlled Environment Agriculture system in an apartment or flat and nobody would know about it until they open the door and found it there. This however is no small undertaking as air conditioners, fans, and other growing equipment can create a significant amount of noise when they are all running simultaneously. A basement, garage or shed may be the best location for setting up a larger indoor garden. As it is these are out of the way and are usually insulated and have access to electricity.

Insulation

Creating or choosing a well-insulated area is a great way of containing the noise of your gardening activities and increasing the overall efficiency of your equipment. Depending on what your growing, or where your growing, it might be preferable that the grow spot has no window, but if it does, you can cover and insulate them while maintaining a pleasant outward appearance.

Savvy, grow room builders sometime construct window boxes which will insulate the windows from condensation, light and sound and provide adequate ventilation to add fresh air or remove humidified air and from the outside the window looks normal complete with a curtain or blind.

Electricity

Specialised indoor grow rooms require electricity, sometimes lots of it. The size of your garden, types of crops and growing system you use will play an important part in determining how much electricity you use. The average domestic electricity circuit should provide you with enough electricity to power a small-scale garden.

However, it is still important to ask yourself: can you safely and easily access the required amount of electricity your garden will need?

A suitable qualified person should be able to redirect electricity from an unused circuit to the garden with little disruption to the rest of the building and restore the existing connections if you decide to dismantle the garden after use.

Irrigation

All indoor plants require some level of irrigation. This means you need easy access to water. For smaller gardens this could be as easy as having a full bucket of water close to your garden. Keep in mind that one large plant can consume as much as four litres of water each day, now consider how practical or easy it will be to carry this much water each day if you have a few similar large plants. If you have difficulty watering your plants, crop quality and yield are likely to suffer.

All the water will also need to drain somewhere. To attain healthy crops you need to over water as this will help to flush away nutrient residues, which may accumulate in your soil or growing medium.

Hydroponic growers need to empty and refill large

nutrient reservoirs at least weekly. Keeping potted plants in hydroponic flood tables allows growers to do this easily by using either using gravity or pumps. If you are growing small-scale hydroponic garden you may decide to use the nutrient run-off in you normal indoor plants or out door garden.

Criminals

Depending on where you live or what you particular circumstances are, portable, expensive and untraceable indoor gardening equipment may appear attractive to some criminals. So it's important that you keep your expensive equipment away from prying eyes. You don't have to go all out with expensive security measure, however the same security measures you would apply to your other valuable items apply here.

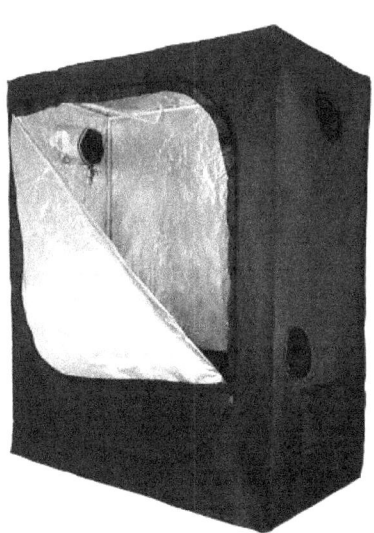

If you live in a confined dwelling and intend to use fans and lights to grow your produce a portable, padded grow room like this may be worth considering.

Chapter 3

How To Select The Ideal Lights For Your Indoor Garden

The right type of grow lighting system can make it easy to grow most type of plants indoors. Any grow light is more effective then no grow light at all, however each type of grow light has it's own pros and cons.

High-Intensity Discharge Or HID Lights

HID lights are the brightest grow lights available. They can be installed in almost any home, greenhouse, or garage as a supplement to the existing light or they can be used as the sole source of lighting for your plants.

Similar in principle to fluorescent lights, these bulbs pass electricity through a glass tube or globe which contains a mixture of gasses. The colour of each type of globe depend on the exact blend of gasses contained in the globe.

HID lights can be twice as efficient as fluorescent lamps; one 400 watt HID globe can emit as much light as the equivalent of 800 watts of fluorescent lighting.

All HID lights can run on normal 120 volt domestic circuit, however they will require special ballets and fixtures.

There Are Two Main Types of HID lamps: high pressure sodium or HPS light and metal halide, also known as MH light. The light emitted by either of these lights is much greater the your typical fluorescent globe.

MH lights emit light which is almost blue in color. It's a cool white light that encourage a compact, leafy growth. Because this lamp will not distort the colors of plants or people it illuminates, this type of lamp is also a great choice for display areas.

Argosun gold halide globe are color corrected to emit a more orange/red light close to natural light. In addition to encouraging compact leafy growth, this light helps boost flowering.

Halide light globes should be replaced yearly.

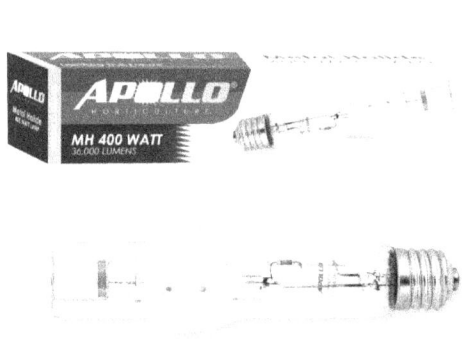

Typical MH Light Globe

HPS Lights last a little longer; theses should be replaced about every eighteen months. The strong light they emit is orange/ red in color which encourages flowering. However. they can also encourage stringy growth unless combined with sunlight or MH lights.

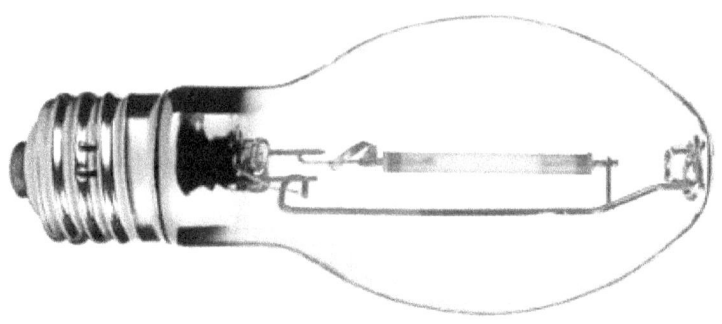

Typical HPS light globe

Grow Light Test Garden Tip: You can combine HPS and MH light globes in single location, but an MH light globe cannot be used in a HPS fixture or vice versa. The HPS ballest contains an igniter which the MH ballests don't. If you have both types of fixture, think about combining both the HPS and MH systems. You can use a conversion bulb which uses metal halide if you only have one fixture, and then switch to a conversion which uses HPS globe to encourage flowering.

Fluorescent High Intensity Lights

Another great choice could be fluorescent high intensity light globes. The fixture looks similar to HID globes, but are more affordable, and warm and cool globes are available which will fit the same ballest. These globes can be chosen according to what you find most appealing.

Standard Fluorescent Lights

Normal fluorescent lights can be among the most affordable ways to get started. You can use second hand shop light fixtures or multi tier glowing carts.

Standard fluorescent lights produce a much less intense light than other alternatives, therefore you may be a little more limited in what you can grow. Standard fluorescent lights may be a good choice if you're just looking to supplement natural light instead of replacing it. Fluorescent tubes come in cool, warm or full-spectrum.

Light from cool tubes has a slight blue tinge to it, warm tube emit a pink/white light, while full-spectrum tube closely look similar to the color of natural daylight.

Full-spectrum tubes tend to cost slightly more than cool or warm tubes, however many indoor growers consider them to be worth the extra money as the color of the light will not distort the color of the plants.

The ends of a fluorescent tube emit less light then the center. Plants that require less light should be beneath the three inches of tube at either end of the fixture.

If you are using the lights approximately sixteen hours a day, then it is a good idea to replace the tubes every eighteen months.

Standard fluorescent light

Led Grow Lights

Older led lights that use hundreds or fractional-watt LED globes usually are not bright enough to be considered as effective replacements for HID lights. However, advancements in LED lights mean that higher quality and newer multiple-watt light may sometimes now yield results exceeding HID lights. Newer LED grown lights are using more electricity which results in increased effectiveness of the technology.

Recent experiments show that providing plants with white LED is also viable because LED colour is achieved by using multiple compounds, therefore it is possible to provide all the wavelengths required with a white LED.

LED lights can cost more to get started, however they generally consume much less electricity and operate at much cooler temperatures than HID lights which means considerable cost savings in electricity down the road.

An example of a high quality LED glow Light

A Formula To Help You calculate How Much Wattage You Require To Grow Your Plants

Once you have decided what kind of lighting you want to use, you need to decide how big a bulb you need for the space you need to light.

First, you need to determine how much space you need to illuminate. Generally, you'll want between twenty and forty watts per square foot. Then divide the wattage of your bulb by twenty. For example (1000 ÷ 20 = 50), now divide the wattage of your globe by forty (1000 ÷ 40 =25).

The answer will give you the highest and lowest extremes of your light intensity range. Depending on your plants and their light requirements with one, one thousand watt system, you can light between twenty five and fifty square feet of interior space.

You can adjust your set-up as you observe how well your plants grow and decrease or increase your lighting intensity.

Adjusting your light intensity can be done by shifting the placement of your plants or the light fixture so that they a further apart or closer together. NEVER attempt to change the light intensity by changing your light globe to one of a higher wattage.

Each globe is designed for a specific wattage. For example a 400 watt bulb cannot safely operate in a 250 watt system.

Chapter 4

Selecting The Right Containers For Your Crops

Luckily, herbs or vegetables aren't too fussed about what kind of pot or container you put them in.
The only basic requirement is that the container is large enough to hold the plant and the that it provided adequate drainage so the excess water can escape.
If you are a beginner, try to use as big a container as practically possible. The reason being is that the bigger pot holds more soil/medium and therefore can retain moisture for longer so you do not have to water the plant as often.

If your using potting mix/soil, try to use containers which are at least ten inches wide and twelve inches deep. Larger flowerpots, half barrels, planters, and large containers such as ten litre buckets all work well.

Some types of vegetables particularly need larger pots to grow in an indoor garden. Most vine crops, such as cucumbers and tomatoes generally prefer containers which are at least twenty inches wide. Peppers prefer containers which are at least fifteen inches wide.
If your container does not already have drainage holes, you will need to add them. This can be done by drilling 5mm holes in the bottom or along the sides of the bottom of the container. Now line the bottom of the pot with either fly mesh or landscaping cloth, this will the the excess water escape, but will allow the soil remain in the pot.

Tall growing plants such as vines will be more productive if they are grown up a support structure inside the pot. For most plants, this can easily be done by inserting a wire cage into the container at when you plat the crop. For trellised plants it is a good idea to use larger and heavier containers as this will reduce the possibility of the plant tipping over.

Indoor Garden Container Materials

Generally, plants in clay pots will need to be watered more often than plants in other types of containers because unglazed clay pots are porous. Also consider the color of the container. Darker colors tend to absorb more heat which could be a problem for some plants depending on you indoor set up.
Avoid, using containers made of treated wood as these may leach out chemical compounds that may be absorbed by your vegetables.

Tip

If you are using potting mix as you grow medium, crops such as corn, carrots, radishes, and spinach, they can be planted directly into a large container, without the need to transplant from as small pot to a larger one. Make sure the potting mix is completely moist before planting by thoroughly watering the container and letting the access water run off for a few hours before planting and always follow the instructions on the back of the seed packets.

Inspect and water the potting mix to insure the potting mix hasn't dried out.

Chapter 5

The Right Gardening Tools

One of the advantages of indoor gardening is you don't need large gardening tools such as forks, rakes, and spades to tend to your garden, which is useful if you don't have the space to store these items.

The are some basic tools which you will need in order to make your gardening experience easier and more productive.

The tools you select will depend on many things such as the size of your garden, what crops you will be planting, and what medium you will be growing your crops in.

The tools listed here are the absolute basic tools you will need.

Pruning Scissors

These are great for pruning vine plants such as tomatoes. They are also great for harvesting herbs such as basil or rosemary.

Spray Pack

The leaves of some plants can become dehydrated especially when planted under hot lights, so spraying them with water can help prevent problems. Spray packs are also useful for spraying plants to prevent parasites.

Small Gardening Set

If you are using soil or potting mix as your planting medium, small gardening sets can be affordable and will make such tasks as transferring potting mix much easier.

A basic gardening set like this has basic tools you need the get started

Watering Can Or Bucket

All plants need water so, if you are planting in a potting mix, having a watering can or some other way to easily water your plants is a good idea.

Chapter 6

The Right Fertilizers

Regardless of which planting medium you choose, all plants need fertilizer to survive and thrive.

For indoor gardeners, creating a living soil, rich in humus and nutrients, is the key to growing vegetables and herbs. The overall fertility and viability of the soil, rather than the application of fertilizers as quick fixes, is at the very heart of indoor gardening.

But, like all gardeners, indoor gardeners have to start somewhere. Your potting may be deficient in certain nutrients, it may not have the best soil structure, or its pH may be too high or too low. Unless you've lucky and can find the perfect potting mix you're going to have to work to make it ideal for gardening.

Indoor gardens especially need fertilizer as the potting mix is usually void of the nutrients plants require selecting the right fertilizer depends on many these. For example if your garden is in an isolated garage, you can mix some manure into the potting mix or feed the plant using a seaweed solution, however these same methods may not be ideal in a home or apartment where people may find the smell offensive.

For convenience water soluble are best. These products are made of chemicals that can easily dissolve or be mixed in water. Simply pour the correct amount in the watering container first, then allow the water from the faucet to mix it.

The chemical based water-soluble fertilizers come in a wide variety of formulas, and their numbers may seem confusing. The first number will be for the percentage of nitrogen, the second phosphorus and the third potassium. Fertilizer formulations will vary according according what each plant needs.

Most of the fertilizers have been packaged for large commercial operations and come in various formulations. The most popular formulation for most indoor crop growers is the 20-20-20 formula. This one will supply general nutrition of most indoor plants. Water-soluble fertilizers also contain the micro nutrients that plant professionals now understand are essential to proper plant growth. Although these nutrients are present in some potting mixes, they are not in all. Even though the micro nutrients are used by plants in minor quantities compared to the three main nutrients, they are important. They include boron, manganese, zinc, copper, molybdenum, and cobalt. The best way to know the individual plant's needs is to plant it and observe how well the plant grows.

Since there are so many plant-nutrition products available, it is important to choose products suitable for you and your plants particular requirements, read product labels carefully, and follow the application rates.

Chapter 7

The Growing Medium

All plants benefit from a good planting medium, with indoor plants this is especially important.

There are various mediums in which you can grow indoor plants, however since the main plant focus of this book is vegetables and herbs, we will be focusing on the two main commonly available mediums which yield the best results, potting/soil mix and hydroponics.

Potting/Soil Mix

Where possible, your potting soil should be mixed according to the particular type of plants you are growing. Seedlings should be grown in a light, moisture-retentive, soil-less mixture.

A good quality indoor potting mix is usually composed of peat moss, perlite and vermiculite. These soil-less mixes absorb moisture very well and resist compaction, but they tend to dry out very quickly. Since they do not contain any nutrients, you must provide your plants with a consistent supply of fertilizer. The main advantage to a soil-less mix is that it is sterile, so there is no chance of introducing pest or disease infestations

Many gardeners add organic components to their indoor growing mix. These might include leaf mold, finished compost, composted peat, or rich garden soil. A growing medium that contains ten to twenty percent organic matter will usually not dry out as easily as a soil-less mix, and will also allow you to introduce beneficial micro-organisms and nutrients.

The most critical consideration when you're purchasing or blending your own potting soil is to ensure that the mix is light enough to provide adequate pore space for air, water and healthy root growth. Month after month of overhead watering, without the benefit of earthworms and weather to aerate the soil, can result in compacted and unhealthy soil around the root zone. To ensure that your plants' roots have the oxygen they need for healthy growth, your potting soil should contain plenty of perlite, vermiculite, or sharp sand. This will allow water to drain freely, and ensure that the soil is at least ten to twenty percent air.

Soil found in your outside garden is not well suited for this purpose, since it's often too heavy and may contain weed seeds and insect pests. If you are purchasing a premixed potting mix, look for a mix that is specific to indoor plants. A good medium should remain loose and drain well, yet contain enough organic matter to hold nutrients and moisture.

Most commercial organic mixes will work well, or you can create your own.
Here is a quick basic recipe for making you own potting mix:

•1 part garden loam or topsoil
•1 part peat moss or mature organic compost
•1 part clean builder's sand or perlite

The organic materials in the mixture will provide structure and the sand will improve drainage. A balanced, slow-release organic fertilizer can also be added to the mix.

Hydroponics

Rather the growing your plants in a potting mix, you may decide to try planting them in a hydroponic medium. This basically means you will be gardening with out soil. With traditional gardening soil holds nutrients and anchors the plant's roots. When you grow crops with hydroponics you directly provide the nutrients an inert medium such as perlite, rock wool, clay pellets, peat moss, or vermiculite may be used purely to hold the roots in place.

Some of the main advantages of growing crops hydroponically include:

• Your plate will grow up to 50% faster since plants can easily access food and water.

• You can use smaller containers as the roots grow throughout the media without becoming root bound.

• Because plants start off in a disease free medium they are less likely to become diseased.

• Should a plant become sick, the problem is usually confined to one plant not multiple plants.

One of the biggest disadvantages with hydroponic gardening is if you nutrient mixture is wrong you could end up harming or at worst killing your plants.

Chapter 8

How To Water Your Indoor Garden

As well as hydrating the plant water serves an important role as a transport medium, allowing nutrients to travel to from the soil to the plant's roots. However too much water can force air from the plant's root zone, which reduces the plant's supply of oxygen.

Unfortunately, there is no simple way of dictating exactly how often a plant should be watered.
The composition of the potting mix and the type of plant pot used play an important role in influencing how efficiently a container retains moisture.

In this chapter I will discuss ways in which should help you determine when you should water you plants and how much you should water them.

How To Evaluate the Moisture Of Your Potting Mix

Plants can be grown in almost any container, however some containers are more suitable than others.

Clay pots are porous and tend to draw moisture from the soil, so your plants will need extra watering. However if the container is glazed, or if it's made of glass, plastic, or any other non-porous material, the container will not absorb moisture you must be careful not to over water.

If your container lacks drainage holes in the bottom, be sure to provide for drainage by adding a layer of gravel in the base of the container where excess water can collect away from the roots and be gradually reabsorbed. Plants grown in closed containers may not need to be watered as frequently.

Learn to spot check a moisture level of your plant's soil. The easiest way to do this is to use a moisture meter which will measure your soil as wet, moist or dry.

Check For Signs Of Over Watering Or Dehydration

When the stems or leaves wilt this could be sign that your plant needs water. Also the potting mix may pull away from the sides of the pot.

Other signs of insufficient water include:

•Slow leaf growth

•Translucent leaves

•Premature leaf shedding

•Leaves yellow and curl

Signs of over watering include:

•Young and old leaves fall at the same time

•Standing water noticed in container under liner

•Leaves have brown soft rotten patches and fail to grow

What To Do If A Plant Is Badly Dehydrated

If the root ball has receded from the sides of its pot, there is no point in trying to water from the top or pot bottom as the water will simply drain through the pot along the sides. It is likely that the plant will be showing considerable stress with leaf/stem wilt or death, but if you have arrived in time, there may still be a chance of saving the plant.

- Using a fork, gently break up the dried-up potting mixture.

- Submerge the pot completely in a bucket filled with water until air bubbles stop rising.

- Prune any dead growth and spray the remaining foliage with a mist of water.

- Move the plant to a cool place and allow any excess water to drain from the pot. In a few hours the plant should begin to recover.

Chapter 9

Hydroponic Basics

What Is Hydroponics?

Hydroponics is a method of growing plants on nutrient rich water based solution. With hydroponics you do not grow plants in soil instead the roots are supported by an inert medium like rock wool, peat moss, or perlite. The point of hydroponics is to promote plant growth by letting the roots come in direct contact with nutrient solution, while also having access to air.

Why Would You Decide To Use A Hydroponic System?
Hydroponics can be an excellent choice especially for the dedicated home gardener as it allows the gardener to precisely control any variables which effect you plant's grow and development. A properly set up hydroponic system can yield much quicker and higher quality results than a soil

based system.

Some of The Advantages Of A Hydroponic System

•The biggest advantage is that if properly set up, plants will grow faster and quicker than with conventional gardening as the plants will not have to work as hard to obtain the nutrients they need.

•You can carefully control the nutrient and PH levels.

•Hydroponic systems generally useless water than soil based systems because the system is enclosed resulting in less evaporation.

Some Of The Disadvantages Of A Hydroponic System

•The biggest disadvantage of a hydroponic system is if you get the growing solution wrong, it could seriously harm your crops or kill them.

•They can be expensive to get started.

•If a pump fails you plants could be dead within hours because the grow medium cannot store water like soil can, so the plants are dependent on a continuous fresh supply of water.

Types of Hydroponic Systems

Hydroponic systems come in various styles and sizes. Some of the best systems are a combination of different type hydroponic system in one unit.

The Drip System

The hydroponic drip system is rather basic. A drip system works by providing a slow stream of nutrient solution to the hydroponics medium. It is recommend using a slow draining medium, such as coconut hair, rock wool, or peat moss.

The major problem with this system is that the drippers have a tendency to clog up. It can be an effective way for growing if you can avoid the clogs that plague this type of system.

Wicking

Is one of the easiest and most affordable hydroponic methods. The basic principle behind wicking is that you have a material like cotton, that is surrounded by a growing medium with one end of the wick material places in the nutrient bath. The solution then travels along the wick to the plants roots.

The system can be further simplified by removing the wicking material completely and just using a medium that can wick nutrients to the roots. This can be achieved suspending the bottom of your medium directly in the solution. For this method it is recommend you use a medium such as perlite or vermiculite.

Aeroponics

Aeroponics is a hydroponics method by where the roots are misted with the nutrient solution while suspended in the air. There are two main methods to get the solution to the exposed roots. The first method uses what's called a pond fogger. The second method involves a fine spray nozzle to mist the roots.

This system can be a great solution for people who would like to grow small herbs such as basil in their kitchen bench as there many compact systems which are now readily available.

If you only intend to grow small quantities of produce an aeroponic system like this may be a good solution

The Deep water Culture System

Deep water Culture, is also known as the reservoir method, is by far the simplest way to grow plants with hydroponics. In a deep water culture system, the roots are suspended in a nutrient solution and an aquarium air pump oxygenates the nutrient solution, this keeps the roots of the plants from drowning. Be sure to prevent light from penetrating your system, as this can encourage algae to grow.

The main benefit to using a Deep water Culture system is that there are no drippers to clog. This makes the deep water culture system an excellent choice for organic hydroponics.

The Nutrient Film System

Nutrient Film Technique, is a hydroponic system where an endless flow of nutrient solution runs over the plants roots. This type of system is on a slight tilt so that the nutrient solution can flow with the force of gravity.

This type of system works well because the roots of a plant can absorb more oxygen from the air than from the nutrient solution itself. Because only the tips of the roots come in contact with the nutrient solution, the plant is able to get more oxygen which encourage faster growth.

The Ebb & Flow System

This ebb and flow system functions by flooding the root zone with the nutrient solution at specific intervals. The nutrient solution then slowly drains back into the reservoir. The pump works by a timer, so the process repeats itself at specific time intervals so that your plants get the required amount of nutrients.

Handy Hints

•The water in you reservoirs should be kept at a temperature of between 65 and 75 degrees Fahrenheit (or 18 to 24 degrees Celsius). This can be done by using a water chiller or heater.

•Change the water in you reservoirs every one to two weeks.

•Follow the feed cycle recommended by the manufacturer of your nutrients.

•If your plants don't look healthy and are either discolored or distorted, then the first thing you should do is check and adjust is the pH. If you determine that the pH is not the problem then flush your system with a solution like Clearex.

•Try using air pump with an air stone connected by flexible tubing as this can help increase the circulation and keep your nutrient solution oxygenated.

•Empty, clean, and sterilize your entire system after you finish growing your crops. Empty the reservoirs and remove any dirt, then run your entire system for about a day with a mixture of non-chlorinated bleach and water. For every gallon of water use 1/8th of a cup of non-chlorine bleach. Then drain your system and flush it thoroughly with clean water to remove any left over bleach.

Chapter 10

How To Grow Vegetables Indoors Without Fancy Equipment

Use The Sunniest Window You Have

Use the sunniest and brightest window in your a apartment or house, most vegetables require plenty of light. Because the light will only be striking one side of the plant, the pot may need to be rotated daily to prevent the plant from leaning on one side. If it lighting still isn't adequate you may still need to add some form of lighting, for this you can try adding fluorescent lights as these a relatively inexpensive. Just don't forget to turn them off at night.

Grow Dwarf veggies

There are many varieties of vegetables you can grow in doors, however you should keep in mind that some types of vegetables may not cope well being grown in containers or grow to enormous proportions. So whatever you decide to plant has to be manageable.

Try to plant smaller varieties of vegetables.

These are after called baby, bush or dwarf veggies or may be described as suitable to container planting.

Unless you are using large containers deep-rooted crops like parsnips can be difficult to grow, however vegetables such as silver beet, baby carrots and radishes can easily be grow indoors.

Plant Taller Crops If You Have The Height

If your the windows are tall enough, you may be able to plant crops such as cucumbers, cherry tomatoes or chilli peppers. Once they have grown to the height of the top window, simple cut the tops off. Baby peas and dwarf beans can grow well indoors, however they will need a stick, wire cage or some other support to hold them up.

Plant An Indoor Salad Garden

Salad vegetables can grow well indoors, and if the window sill is well lit, they will even grow through the winter. They can be planted closer together than outdoor veggies, and taste ever better when they are picked young.

Sow them regularly so you have a constant supply. By doing this you can start every meal with a fresh salad which will add additional nutrition to your meals everyday.

Use Your Plants as Decoration For Your Home

Instead of planting regular varieties that can be purchase at any supermarket, try to plant varieties which produce attractive flowers or unusually colored crops such as purple carrots or yellow tomatoes. These type of crops can serve as edible house decoration.

If you decide to use your plants as edible decoration you will need attractive or unusual containers and pots to display them in try using nice, glazed or plain terracotta pots or if you like recycling and would like to achieve a rustic look to you home you can use items such as old metal buckets, old watering cans and even food cans. If you containers don't have any drainage holes make sure you drill some so your crops don't get water logged.

Pollinate Your Crops

It is unlikely that your fruiting plants such as beans and tomatoes will encounter wind or bees to help pollinate them. You may need to lend a helping hand. To do this use a soft paintbrush to transfer pollen from one plant to another.

Don't Over water Your Plants

Don't over water your plants as this will deprive their root of oxygen, however use a quality liquid feed to feed them well and don't forget to add drip trays under your containers to protect you carpets and floors.

Grow Mushrooms

Mushrooms don't need light, try to grow them in a poorly lit room or cupboard. Use compost with pre-seeded mushroom spawn or use a log which has been specially prepared if you want to grow different varieties.

Chapter 11

How To Grow Herbs Indoors Without Fancy Equipment

Many types of herbs can be grown indoors with out the aid of extra lighting. Here are some of the best herbs you can grow on windowsills and near natural light sources such as glass doors and windows and some handy hints you can implement to keep them healthy.

Cutting And Rooting

How To Plant Using a Cutting

Many herbs including oregano, rosemary, thyme and sage are ideally propagated for indoor growing by taking a cut from an existing outdoor plant. To do this, cut off a 4-inch section, measured back from the tip. Strip off the lower leaves and stick the stem into moist, potting mix. To ensure good humidity, cover with glass or clear plastic, and keep the potting mix moist.

Light, Water And Temperature

Only plant herbs in a container which can provide adequate drainage, while most herbs like to be well watered, so it can be quite easy to over water them.

Only water when the surface of the potting mix feels dry and add sand or vermiculite to the potting mixture to encourage proper drainage.

Learn to fine tune water, light, and temperature requirements of your plant. A plant in a terracotta pot in a south-facing window will need more water than one in a plastic pot in an east or west facing window. If the light is low, keep the temperature low.

Hold back on the water and fertilizer through winter, but when the days start getting longer, feed them with liquid seaweed or compost. Even potted soil gets compacted as you water it, so cultivate it with a little fork, then top off the surface of the soil with compost and mulch.

Spring is usually a great month for indoor plants because of all the bright light. During summer make sure the plants are well watered as they can quickly become dehydrated in the heat under a window.

Here's a list of herbs you can grow indoors affordably using basic gardening equipment.

Basil: Should be planted from seed. Place the pot in a south facing window as they like plenty of warmth and sun.

Bay: Can be grown in containers all year round. Ideally pots should be place under windows that face east or west and insure they positioned somewhere where the air can circulate freely.

Chervil: Sow chervil seeds in late summer. It grows well in low light but grows best at a temperatures of between 65 to 70 degrees Fahrenheit (18 to 24 degrees Celsius).

Oregano: Can be sown from seed, but is ideally grown using a cutting. Position the pot in a south-facing window.

Parsley: This herb can be sown from seed. Parsley likes full sun, but will slowly grow in an east or west, facing window.

Rosemary: Plant rosemary using a cutting and keep it in moist potting mix until it roots. It grows best in a sunny south-facing window.

Sage: Plant sage using a cutting. Sage will tolerate dry, indoor air well, but will need the strong sun it will get in a south-facing window.

Thyme: Plant thyme using a cutting. Thyme prefers full sun but will also grow in an east, or west, facing window.

www.ingramcontent.com/pod-product-compliance
Lightning Source LLC
Chambersburg PA
CBHW070715180526
45167CB00004B/1483